YOUR KNOWLEDGE HAS VALUE

- We will publish your bachelor's and master's thesis, essays and papers

- Your own eBook and book - sold worldwide in all relevant shops

- Earn money with each sale

Upload your text at www.GRIN.com and publish for free

Bibliographic information published by the German National Library:

The German National Library lists this publication in the National Bibliography; detailed bibliographic data are available on the Internet at http://dnb.dnb.de .

Imprint:

Copyright © 2018 GRIN Verlag
Print and binding: Books on Demand GmbH, Norderstedt Germany
ISBN: 9783668757127

This book at GRIN:

https://www.grin.com/document/432094

Faizan Danish Khaleel

Natural products and its scope and applications

GRIN Verlag

Inhaltsverzeichnis

1. Introduction

A natural product is a chemical compound/ substance, generated by living organisms found in nature. In the broad sense, it includes any substance produced by life. It has a pharmacological/ biological activity which is used in pharmaceutical drug discovery and design. A natural product can be prepared by total synthesis (Fig 1.1).

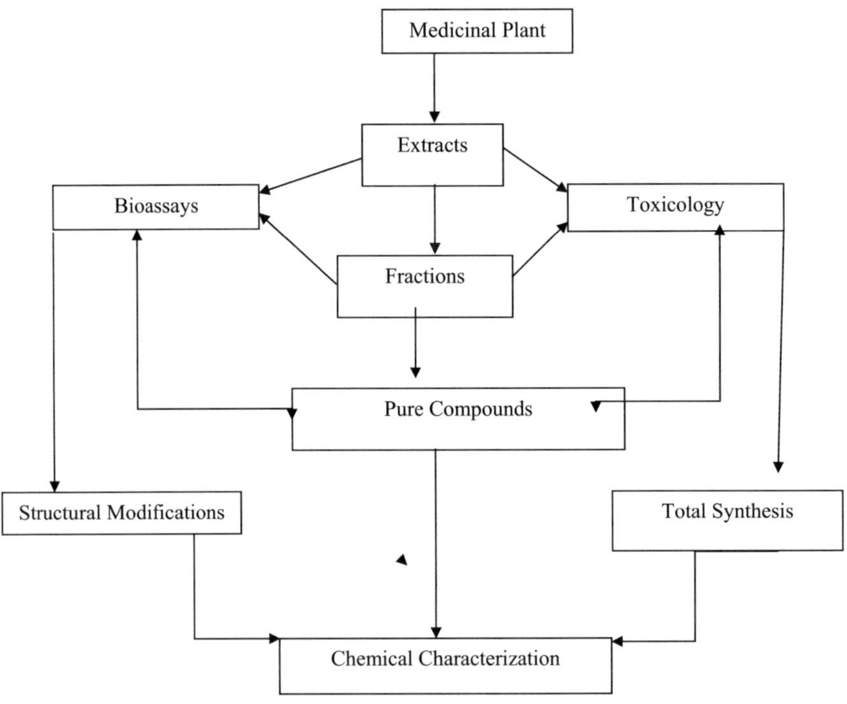

Fig. 1.1

Natural products have played a vital role in the development of organic chemistry (purified organic compounds isolated from natural sources) by providing hectic synthetic targets. The term, Natural product has been extended to commercial purposes like cosmetics, dietary supplements and food produced without any artificial ingredient.

Natural products are the active components of traditional and modern medicines. As the structural diversity of natural products helps in the chemical synthesis, and synthetic analogs can be prepared with improved potency and safety, these are often used as the starting points of drug discovery. In the field of organic chemistry, natural products are often defined as primary and secondary metabolites. Another definition limiting natural products to secondary metabolites is used in the fields of medicinal chemistry and pharmacognosy[23].

At present, there are more than 120 chemical substances obtained from plants that are considered as important drugs which are currently used in a number of countries of the world. These chemical substances are given in the Table (1.1) shown below[18].

Table 1.1

Drug/Chemical	Action/ Clinical use	Plant source
Acetyldigoxin	Cardiotonic	Digitalis lanata
Adoniside	Cardiotonic	Adonis vernalis
Aesein	Anti-inflammatory	Aesculus hippocastanum
Aesculetin	Antidysentery	Fraxinus rhynchophylla
Agrimophol	Anthelmintic	Agrimonia eupatoria
Allyl isothiocyanate	Rubefacient	Brassica nigra
Allantoin	Vulneray	Several plants
Anabesine	Skeletal muscle relaxant	Anabasis sphylla
Andrographolide	Bacillary dysentery	Andrographis paniculata
Anisodamine	Anticholinergic	Anisodus tanguticus
Anisodine	Anticholinergic	Anisodus tanguticus
Areceline	Anthelmintic	Areca catechu
Asiaticoside	Vulnerary	Centella asiatica
Atropine	Anticholinergic	Atropa belladonna
Berberine	Bacillary dysentery	Berberis vulgaris
Bergenin	Antitussive	Ardisia japonica
Bromelain	Anti-inflammatory, proteolytic agent	Ananas comosus
Betulinic acid	Anticancerous	Betula alba
Borneal	Antipyretic, Analgesic, Anti-inflammatory	Several plants
Caffeine	CNS stimulant	Camellia sinensis
Camphor	Rubefacient	Cinnamomum camphora
Camptothecin	Anticancerous	Comptotheca acuminate
(+)-Catechin	Haemostatic	Potentilla fragaroides
Chymopapain	Proteolytic, mucolytic	Carica papaya
Cissampeline	Skeletal muscle relaxant	Cissampelos pareira
Cocaine	Local anaesthetic	Erythroxylum coca
Codeine	Analgesic, Antitussive	Papaver somniferum
Colchiceine amide	Antitumor agent	Calchicum autumnale

Colchicine	Antitumour agent, Antigout, Anti-inflammatory	Colchicum autumnale
Convallotoxin	Cardiotonic	Convallaria majalis
Curcumin	Choleretic	Curcuma longa
Cynarin	Choleretic	Cynara scolymus
Danthron	Laxative	Cassia spp.
Demecolcine	Antitumor agent	Colchicum autumnale
Deserpidine	Antihypertensive; tranquilizer	Rauvolfia canescens
Deslanoside	Cardiotonic	Digitalis lanata
L- dopa	Anti- parkinsonism	Mucuna sp.
Digitalin	Cardiotonic	Digitalis purpurea
Digitoxin	Cardiotonic	Digitalis purpurea
Digoxin	Cardiotonic	Digitalis lanata
Emetine	Amoebicide; emetic	Cephaelis ipecacuanha
Ephedrine	Sympathomimetic	Ephedra sinica
Etoposide	Antitumour agent	Podophyllum peltatum
Galathamine	Cholinesterase inhibitor	Lycoris Squamigera
Gitalin	Cardiotonic	Doigitalis purpurea
Glaucaroubin	Amoebicide	Simarouba glauca.
Glaucine	Antitussive	Glaucium flavum
Glasiovine	Antidepressant	Octea glaziovii
Glycyrrhizin	Sweetener	Glycrrhiza glabra
Gossypol	Male contraceptive	Gossypium spp.
Hemsleyadin	Bacillary dysentery	Helmsleya amabilis
Hesperidin	Capillary fragility	Citrus species
Hydrastine	Hemostatic; astringent	Hydrastis Canadensis
Hyoscamine	Anticholinergic	Hyoscamus niger
Irinotecan	Anticancer; antitumour agent	Camptotheca acuminate
Kainic acid	Ascaricide	Digenea simplex
Kawain	Tranquilizer	Piper methysicum
Khellin	Bronchodilator	Ammi visnaga
Lanatoside, A,B,C	Cardiotonic	Digitalis lanata
Lapachol	Anticancer, antitumor	Tabebuia sp.
Lobeline	Smoking deterrent, respiratory stimulant	Lobelia inflanta
Menthol	Rubefacient	Mentha species
Methyl salicylate	Rubefacient	Gaultheria procumbens
Monocrotaline	Antitumour agent	Crotolaria sessiliflora
Morphine	Analgesic	Papaver somniferum
Neoandrographolide	Bacillary dysentery	Andrographis paniculata
Noscapine	Antitussive	Papaver soniferum
Nicotine	Insecticide	Nicotiana tobacum
Nordihydroguariaretic	Antioxidant	Larrea divaricata
Quabain	Cardiotonic	Strophanthus gratus
Pachycarpine	Oxytocic	Sophora pschycarpa
Palmatine	Antipyretic;detoxicant	Coptis japonica
Papain	Proteolytic; Mucolytic	Carica papaya
Papaverine	Sympatholytic musculotropic	Papaver somniferum
Phyllodulcin	Sweetener	Hydrangea macrrophylla

Physostigmine	Cholenesterase inhibitor	Physostigma venenosum
Picrotoxin	Analeptic	Anamirta cocculus
Pilocarpine	Parasympathomimetic	Pilocarpus jobarandi
Pinitol	Expectorant	Several plants
Podophyllotoxin	Condylomata acuminate (tropical antiviral agent)	Podophyllum peltatum
Protoveratrines A & B	Antihypertensive	Veratrum album
Pseudoephedrine	Sympathomimetic	Ephedra sinica
Pseudoephedrine	Sympathomimetic	Ephedra sinica
Quisqualic acid	Anthelmintic	Quisqualis indica
Quinidine	Antiarrhythmic	Cinchona ledgeriana
Quinine	Antimalaric	Cinchona ledgeriana
Rescinnamine	Antihypertensive; tranquilizer	Rauvolfia serpentine
Reserpine	Antihypertensive; tranquilizer	Rauvolfia serpentine
Rhomitoxine	Antihypertensive	Rhododendron serpentine
Rorifone	Antitussive	Rorippa indica
Rotenone	Piscicide	Lonchocarpus nicon
Rotundine	Analgesic; sedative	Stephania sinica
Rutin	Capillary fragility	Citrus species
Salicin	Analgesic	Salix alba
Sanguinarine	Dental plaque inhibitor	Sanguinaria Canadensis
Santonin	Ascaricide	Artemisia maritime
Scillarin A	Cardiotonic	Urginea maritime
Scopolamine	Sedative	Datura metel
Sennosides A & B	Laxative	Cassia spp.
Silymarin	Antihepatotoxic	Silybum mariamum
Sparteine	Oxytocic	Cytisus Scoparius
Theobromine	Diuretic; Bronchodilator	Theobroma cacao
Trichosanthin	Abortifacient	Thymus vulgaris
Tubocurarine	Skeletal muscle relaxant	Chondodendron tomentosum
Vasicine	Cerebral stimulant	Vinea minor
Vincamine	Cerebral stimulant	Vinea minor
Xanthotoxin	Leukoderma, Vitiligo	Ammi majus
Yohimbine	Aphrodisiac	Pausinystalia yohimbe

Some of the drugs/ chemicals shown in Table 1.1 are still sold as plant based drugs, requiring the processing of plant material. Others have been synthesized in laboratories and no plant materials are used in the manufacture of drugs. A good example of this is the plant chemical quinine, which was extracted from the bark of *Cinchona ledgeriana* tree and processed into pill to treat malaria. Nowadays, all quinine drugs available are manufactured chemically without the use of tree bark.

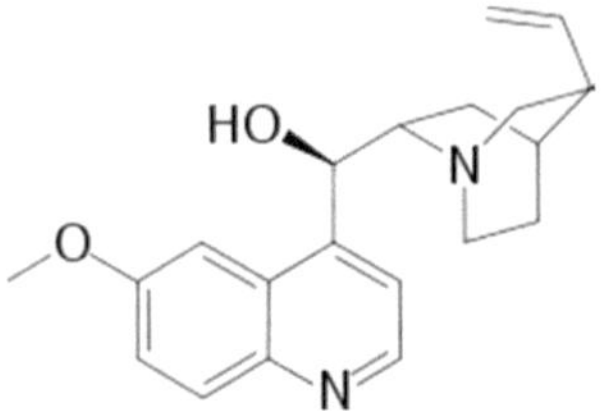

2. Sources of natural products

Several drugs are derived from various naturally occurring medicinal sources. These can be broadly divided into four catagories:

Sources of natural products:

1.2.1 Plant sources

1.2.2 Animal sources

1.2.3 Microbial sources

1.2.4 Marine sources

2.1 Natural products from Plant sources

The use of plants as medicines has a long history in the treatment of various diseases. The earliest known records for the use of such plants are from Mesopotamia in 2600 B.C.[24] Till date, 35,000- 70,000 plant species have been screened for their medicinal uses[13] like taxol, camptothecin, morphine and quinine. The first two are widely used as anticancer drugs, while the remaining are analgesic and antimalarial agents, respectively.

Morphine

Camptothecin

2.2 Natural products from Animal sources

Animals have also been a source of some interesting compounds that can be used as drugs. Epibatidine, obtained from the skin of an Ecuadorian poison frog, is ten times more potent than morphine[31]. Venoms and toxins from animals have played a significant role in designing a multitude of cures for several diseases. Teprotide extracted from a Brazilian viper, has led to the development of cilazapril and captopril, which are effective against hypertension.

Epibatidine

Captopril

2.3 Natural products from Microorganisms

Microorganisms as a source of potential drug candidates were not explored until the discovery of penicillin in 1929. Since then, a large number of terrestrial and marine microorganisms have been screened for drug discovery. Microorganisms have a wide variety of potentially active substances and have led to the discovery of antibacterial agents like cephalosporins, and anticancer agents like epirubicin[11].

Cephalosporin

Epirubicin

2.4 Natural products from Marine organisms

The first active compounds to be isolated from marine species were spongouridine and spongothymidine.These compounds are nucleotides and show great potential as anticancer and antiviral agents. Their discovery led to an extensive research to identify novel drug candidates from marine sources. About 70% of the earth's surface is covered by the oceans, providing significant biodiversity for exploration for drug sources. Many marine organisms have asedentary lifestyle, and thereby synthesize many complex and extremely potent chemicals[15]. These chemicals can serve as possible remedies for various ailments, especially cancer. One such example is discodermolide, isolated from marine sponge, discodermia dissolute, which has similar mode of action to that of paclitaxol and possesses a strong antitumor activity. It also exhibits better water solubility as compared to paclitaxol. A combination therapy of the two drugs has led to reduced tumor growth in certain cancers.

Discodermolide

3. Applications of natural products

3.1 Drugs and Drug leads

Quinine was first used for malaria and acts both as a drug and a drug lead. Same is the case with vincristine and vinblastine for a number of years. They were used to treat cancers mainly as semi- synthetic derivatives. The latest example of a natural product acting as a drug lead is the venom of a viper. Based on the structure of a peptide isolated from this, majority of the 'prils' (enalapril) were developed.

Vincristine
Vinblastine

3.2 Cosmetics and perfumes

The materials used in perfumes, cosmetics are the volatile oils of natural origin. Though it is a multi- billion dollar industry, yet the importance of natural products is un- noticed.

3.3 Defence

Though there is no practical evidence, it is believed that natural products can play a great role in defence, war- fare, rockets etc. while some can act by causing allergy, others can help in ignition processes and even some can have propellant action. Rockets for Diwali celebrations were being manufactured in India much before the space technology development.

3.4 Pesticides

The best and the most famous example of this is Azadirachtin, the source of which is the Neem tree. Bio- pesticides have caught the attention of the whole world after this discovery.

Azadirachtin

4. Medicinal plants

The development of traditional medicinal system including plants as a means of therapy back to the middle Paleolithic age (some 60,000 years ago). Ethno- medicine (use of plant by humans as a medicine) is a highly modified approach to drug discovery. It involves observation, description in experimental investigation/screening for possible medicinal and biological properties of indigenous drugs. It is based on chemistry, botany, pharmacology, physiology, biochemistry, anthropology, history and archaeology that contribute to the discovery of natural products with medicinal activity[4]. According to W.H.O more than 60% of the world population uses ethno-medicine as part of their basic health care.

It is well established fact that plants have been used by humans for thousands of years, so it could, therefore, be inferred / expected that any bioactive compounds obtained from such plants to have low human toxicity. Therefore, the goals of using plants for therapeutic use as drugs can be summed up as under:-

1. To isolate and characterize bioactive compounds for use as drugs like the cardiac drugs, morphine, taxol etc.
2. To produce bioactive compounds of known structures as lead compounds for semi-synthesis, to produce pharmaceuticals that may show lower toxicity like metformin, verapamil and amiodarone.
3. To use agent as pharmacological tools.
4. To use the whole plant / part of it as a herbal remedy like garlic, bitter melon etc.

5. Ayurveda and Traditional medicines

Ayurveda is one of the most ancient and living tradition widely practiced in Sri Lanka, India and other countries and has a sound experimental and philosophical basis[12]. There are over 700 herbs whose description is given in Charaksamhita, Sushrutsumhita[6] and Atharvaveda. Indian health care system comprises of medical pluralism and Ayurveda remains dominant still compared to modern medicine, particularly for the treatment of chronic diseases[34].

A glimpse of Ayurvedic heritage can be obtained from the selection of top twenty Ayurvedic drugs for each of which, we have given one indicative key reference[14,29] Table 1.2.

Table 1.2

S.No.	Sanskrit name	Botanical name	Main activity	Key reference
1	Amalaki	Phyllanthus emblica	Rasayana	26
2	Ashwagandha	Withania somnifera	Immunomodulatory	21
3	Bhallataka	Semecarpus anacardium	Antiarthritic	27
4	Bilva	Aegle mermelos	Antidiarrhoeal	30
5	Chandan	Santalum album	Antiviral	5
6	Chitraka	Plumbago zeylancia	Antitumor	17
7	Dadima	Punica granatum	Anti-diarrhoeal	25
8	Eranda	Ricinus communis	Hepatoprotective	32
9	Guduchi	Tinospora cordifolia	Immunomodulatory	20
10	Haridra	Curcuma longa	Antimicrobial	3
11	Haritaki	Terminalia chebula	Hypolipidermic	28
12	Manjishtha	Rubia cordifolia	Antioxidant	33
13	Maricha	Piper nigrum	Bio- enhancer	16
14	Neem	Azadirachta indica	Anti- diabetic	9
15	Pippali	Piper longum	Bio- enhancer	2
16	Sariva	Hemeidesmus indicus	Anti- ulcer	1
17	Shunthi	Zingiber officinale	Anti- emetic	7
18	Vacha	Acorus calamus	Psycho-tropic	8
19	Vidanga	Embelia ribes	Anti-fertility	21
20	Yashtimadhu	Glycyrrhiza glabra	Anti- ulcer	18

6. Positive contributions from traditional medicines

Adatoda vasica is a well known anti- aesthamatic but vasicine, the bioactive alkaloid from this plant cannot fit the category of a drug. Similarly, brahmi (bacopa monnieri) has been used since centuries as a memory enhancer but, baccosides isolated from this plant does not provide similar results as the whole crude drug itself. Triphala has been used for cleansing of bowels and as a general well being tonic, however, the individual constituents of the plant used in its formulation i,e. polyphenolics and tannins fail to exert the benefits that the whole formulation produces. The above mentioned examples suggest that the traditional Ayurvedicplants have a great potential in their potent form clinically and not as individual components. Therfore, not only efforts for the better pharmaceutics should be directed to yield better formulation but also the standardization of herbal medicinal products needs improvement and a stringent control for the allopathic medicinal system[10].

1.3 Natural products from Indian medicinal plants and their biological activity[10]

Plants/herbal formulation	Chemical constituents	Biological activity/Indication	Market/Traditional formulation
Adhatada	Vasicine, Vasicinone	Bronchodilator	Diakot, Koflet
Euphorbia vasica	Flavonoids	Piles	Thank God
Cassia ssp	Sennosides	Constipation	Kayamchurna
Bacopa monnieri	Baccosides	Memory enhancer	Mental, Himalaya Bramhmi
Tylophora indica	Tylophorine	Bronchodilator	Fizzle, Vasa fort
Triphala	Polyphenolics, Tannins	Bowel-cleanser, General tonic	Triphala
Holarrhena antidysentrica	Conessine	Antiamoebic	Kutajarista
Asparagus adscendens	Asparanin A$B, Sarasapogenin	Fertility enhancer	Spermon
A. racemosus	Shatavarin	Galactogogue tonic	Geriforte
Ocimum santum	Monoterpenes / Sesquiterpenes	Respiratory diseases / Immunomodulatory	Kofostal syrup / Curill
Glycyrrhiza glabra	Glycrrhizin	Anti-ulcer, anti-tussive	kofex
Aloe vera	Aloin	Demulcent	Clarina
Tribulus terrestris	Protodioscin	Diuretic, aphrodisiac	Goksnura
Withania somnifera	Withanolides	Immunomodulatory	Ashwagandharista
Embelia ribes	Embelia	Antifertility	Pipalayadi yoga
Pterocarpus marsupium	Liquiritigenin / Isoliquiritigenin	Antidiabetic	Diabecon
Tinospora cordifolia	Tinosporic acid / Cordifolioside	Immunomodulatory	Himalaya guduchi
Aegle marmelos	Aegelin, marmelosin	Bowel diseases	Diarex
Phyllanthus emblica	Polyphenolics, Tannins	Antioxidant	Chyavanprasha
Centella asiatica	Asiaticiside	Memory enhancer	Mentat
Garcinia cambogia	Hydroxy citric acid	Antiobesity	Bioslim, Ayusilm

Psoralea corylifolia	Psoralen	Vitiligo	Pigmento
Areca catechu	Tannins	Antiobesity, antitussive	Koflet, bioslim
Gmelina arborea	Arboreol	Tonic, stomachic	Chyavanprasha
Achyranthes aspera	Achyranthine	Diuretic	Cystone
Antethum graveolens	Anethole	Digestive, carminative	Bonnisan
Argyreia nervosa	Alkaloids	Aphrodisiac, fertility enhancer	Confido
Vitex negundo	Flavonoids	Anti-inflammatory	Himcolin
Bahuhinia variegate	Boeravinones	Diarrhea, piles	Pilex
Boerrhavia diffusa	Boeravinones	Hepatoprotective	Liv52
Cyperus rotendus	Monoterpenes Sesquiterpenes	Antibacterial Antipyretic	Himpyrin
Evovulus alsinoides	Flavenoids	Bitter, tonic	Anxocare
Symplocos	Alkaloids	Gynaecological disorders	Evecare
Eugenia gambolena	Anthocyanins	Antidiabetic	Diabecon

References

1. Anup, A., Jegadeesan, M., (2003). Biochemical studies on the antiulcerogenic potential of Hemidesmus Indicus R. Br. Var. Indicus. *J. Ethnopharmacol.*, 84, 149.
2. Atal, C. K., Zutshi, U., Rao, P. G., (1981). Scientific evidence on the role of ayurvedic herbs on bioavailability of drugs. *J. Ethnopharmacol.*, 4, 229- 232.
3. Banerjee, A., Nigam, S.S., (1978). Antimicrobial efficacy of the essential oil of *Curcuma longa. Indian J. Med. Res.*, 68, 864.
4. Bannerman, R. H. O., Burton, J., Chen, W. C., (1983). Traditional medicine and health care coverage. A reader for health administrators and practioners, Geneva: World health Organisation.
5. Benencia, F., Courreges, M. C., (1999). Antiviral activity of Sandalwood oil against herpes simplex viruses- 1 and 2. *Phytomedicine, 6*, 119-123.
6. Bhagavan das, Sharma, B. K., (2001). Charak Samhita, Chaukhamba (Sanskrit series office, Varanasi, India) 7th *edn.*
7. Cannel, R. J. P., (1998). Follow- up of natural product isolation. *Methods in Biotechnol.* 4, 425- 463.
8. Cannel, R. J. P., (1998). How to approach the isolation of a natural product. *Methods in Biotechnol.* 4, 1- 5.
9. Chattaopadhyaya, R., (1993). Possible mechanism of anti- hyperglycemic effect of Azardirachta indica leaf extract- part I. *Fitoterpia*, 64, 332.
10. Chen, Y., Liang, X., (2001). Structural chemistry and biological activities of natural products. J. *Chinese Herb. Med.*, 1, 215- 321.
11. Chin, Y. W., Balunas, M. J., Chai, H. B., Kinghom, A. D., (2006). Drug discovery from natural sources. *J. Aaps.*, 8, 239- 253.
12. Chopra, A., Doiphode, V., (2002). Ayurvedic medicine: Core concept, therapeutic principles, and current relevanc. *Med. Clin. North Am.*, 86, 75- 89.
13. Dev, S., (1999). Ancient modern concordance in Ayurvedic plants, some examples, environmental health prospective. 107, 783- 789.
14. Devasagayam, T. P. A., Sainis, K. B., (2002). Immune system and antioxidants: especially those derived from Indian medicinal plants. *Indian J. Exp. Biol.*, 40, 639- 655.
15. Haefner, B., (2003). Drugs from the deep: marine natural products as drug candidates. *Drug Discovery today*, 8, 536- 544.
16. Indian Herbal Pharmacopoeia, Indian Drug Manufacturing Association, (2002). p.317 (2002).
17. Kavimani, S., (1996). Antitumour activity of plumbagin against Dalton's ascetic lymphoma. *Indian J. Pharm. Sci.*, 58, 194- 196.
18. Leslie Taylor, N. D., (2000). "Plant based drugs and Medicines" Rain Tree Nutrition.
19. Mabry, T. J. (2001). Selected topics from forty years of natural products research: Flavonoids, antiviral proteins and neurotoxic nonprotein amino acids. *J. Nat. Products.*, 64 (12), 1596-1604.

20. Manjiarekar, P. N., Jolly, C. I., Narayana, S., (1982). Comparative studies of immune-modulatory activity of T. Cordifolia and T. chinensis. *Fitoterpia,* 71, 254- 257.

21. Mohammed Zauddin, (1996). Study of the Immunomodulatory effects of Ashwagandha. *J. Ethnopharmacol.,* 50, 69-78.

22. Nakanishi, K., (1999). An historical prospective of natural products chemistry. *Comprehensive natural products chemistry,* 1, 23-40.

23. Newman, D. J., Cragg, G. M., 2007. Natural products as sources of new drug over the last 25 years. *J. of Nat. Prod.,* 70, 461- 477.

24. Newman, D. J., Cragg, G. M., Snader, K. M., (2000). The influence of natural products upon drug discovery. *Nat. Prod. Res.,* 17, 215- 234.

25. Pillai, N. R., (1992). Antidiarrhoeal activity of Punica granatum in experimental animals. *Int. J. Pharmacog.,* 30, 201- 204.

26. Rege, N., Thatte, U., Dahanukar, S., (1999). Adaptogenic property of six rasayana herbs used in Ayurvedic medicine. *Phytother. Res.,* 13, 275.

27. Saraf, M. N., Ghooi, R. B., Patwardhan, B. K., (1989). Studies on the mechanism of action of Semecarpus anacardium in rheumatoid arthritis". *J. Ethnopharmacol.,* 25, 159-164.

28. Shaila, H. P., (1998). Hypolipidemic activity of three indigenous drugs in experimentally induced atherosclerosis. *Int. J. Cardiol.,* 67, 119.

29. Sharma, P. C., Yelne, M. B., Dennis, T. J., (2001). Database born Medicinal Plants used in Ayurveda, (Central Council for Research in Ayurveda and Siddha, New Delhi), Vols 1-3.

30. Shoba, F. G., Thomas, M., (2001). Study of antidiarrhoeal activity in Castor oil induced diarrhea. *J. Ethnopharmacol.,* 76, 1.

31. Spande, T. F., Garraffo, H. M., Edwards, M. W., Yeh, H. J. C., Pannell, L., and Daly, J. W., (1992). Epibatidine- a novel (Chloropyridyl) Azabicycloheptane with potent analgesic activity from an Ecuadorian poison frog. *J. of Am. Chem. Soc.,* 114, 3475-3478.

32. Thara- Walanave, M., (1999). Effects of Castor oil on lipid metabolism in rats. *Biosci. Biotechnol. Biochem.* 3, 595- 597.

33. Tripathi, Y. B., Sharma, M., (1998). Comparison of the anti-oxidant action of the alcoholic extract of Rubia cordifolia with rubiadin". *Indian J. Biochem.,* 35, 319.

34. Waxler- Morrison, N. E., (1988). Plural Medicine in India and Sri Lanka: do ayurvedic and Western medical practices differ. *Soc. Sci. Med.,* 27, 531- 544.